CW01500287

Wildlife Wong

and the

Orangutan

by
Dr Sarah Pye

Wildlife Wong and the Orangutan
May 2021

ISBN: 978-0-6451543-0-6 (paperback)
ISBN: 978-0-6451543-1-3 (ebook)

Published by:

Estralita Publishing
ABN: 86 230 144 690
P.O. Box 288
Buddina, QLD, 4575
⊕ www.sarahrpye.com

Pencil sketch illustrator: *Ali Beck*
Cover design: *Gram Telen*
Layout design: *Gram Telen*
Wildlife Wong cartoon illustrator: *Isuru Pltawala*
Cover author photo: *Amber Grant*

A catalogue record for this work is available from the National Library of Australia

Check out what other kids think about this book...

*"I loved the funny stories about what the orangutans get up to in the rescue centre, they certainly are naughty.
I would love to visit some of the interesting places in this book like the rainforests in Borneo.
I learnt lots about primates and how similar they are to us. I wish we could climb and hang from trees as well as they can!"*

Samuel, age 10

"This book is great for readers who are interested in wildlife and conservation. It tells the story of Wong and his love for unusual animals. I especially like how he has worked to save orangutans from captivity. I also learnt lots of nature facts. 5 out of 5!!!"

Issie Thorpe, age 9

"I really enjoyed this book. On each page I could find at least one fact that I didn't know, and each was very interesting. I also liked the way the facts were mixed in with the story."

Alex Messetter, age 12

"We loved this book!
Many of us loved the funny stories
about the orangutans most of all.
Others loved learning so many
new facts about orangutans and
conservation most of all.
We think this book is excellent for all readers
who love animals and the environment. It has
inspired us to want to be like
Dr Wong and save our world's wildlife."

Ellie Eakins and class 4J, age 9-10

Author Sarah in the jungle!

My name is Sarah and I am Wildlife Wong's friend. Wong's whole name is Wong Siew Te. He is an **ecologist** which means he is a scientist who studies how animals and plants live together. He lives on an island called Borneo, on the edge of a thick jungle. Borneo is home to over a thousand different species. Some of them are **endemic** (which means they are not found anywhere else). It's a perfect place for an ecologist to live because there are many different animals to observe. I want to tell you a story about Wong and the orangutans he has met. Sometimes it will be funny, and other times sad.

But I am getting ahead of myself. Let's start at the beginning!

This book is different

This book includes cool stuff to do so you can be a scientist just like my friend Wong. It is the second book in a series, but don't worry... you don't have to read the first one first!

Would you like to make your own book? If you go to my website, www.sarahrpye.com, you can download free pages for your own Nature Journal and a template for making a cool cover. Does that sound like fun? Then let's get on with the story...

Wildlife Wong meets an orangutan

The first time Wong saw an orangutan he was about your age. He grew up in a place called Penang, in a country called Malaysia. One day, his parents took him to the botanical gardens. A botanical garden is a place of bushland where people can experience nature. Sometimes it has examples of **introduced** trees and plants (which means they wouldn't grow there naturally). When Wong was growing up, the Penang Botanical Garden also included caged animals.

Can you see Penang on this map? Hint: Penang is an island on the west coast of West Malaysia.

This is Malaysia

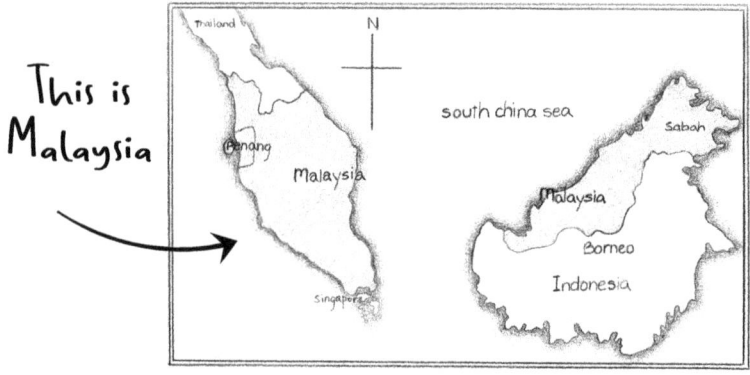

Can you also see that Malaysia is divided into two parts? The first part, West Malaysia is sometimes also called Peninsular Malaysia. **Peninsular** is the name given to land that juts out into the ocean. The northern end of the peninsular joins Malaysia to the country of Thailand (home to elephants, Buddhist temples and Pad Thai). At the very bottom of the peninsular is the island nation of Singapore (home to amazing treelike buildings, and Singapore noodles!). The other part of Malaysia is on the island of Borneo. This is where Wong lives now.

A very, very long time ago, about 2.6 million years ago, the water level was 200 metres below where it is today, and all these islands and countries were attached. At that time, the area of land was called Sundaland. If the sea level continues to rise, there might be even more islands in the future.

A boy who loved animals

When Wong was little, his dad was a **tailor**, which is someone who sews clothes. Wong called his dad Apak (which sounds like ay-pack) and his mum Ah Ji (which sounds like R-gee). Apak worked very hard from early in the morning until late at night. Wong's uncle, aunt and cousins lived with them. There were lots of clothes to make, and all the family was expected to help. After school, Wong swept the floor around his sisters' sewing machines so they could work faster.

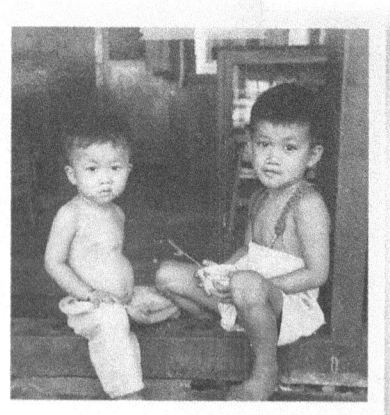

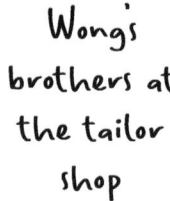

Wong's brothers at the tailor shop

PHOTOGRAPH: © WONG FAMILY

One weekend, Apak thought his family had done such a good job that they deserved a day off. While Ah Ji packed left-over snacks into a bag, Apak checked the oil in the car. It wasn't far to the botanical gardens, but he wanted to make sure the car would make it.

This was before people had to wear seatbelts. Some of Wong's brothers, sisters and cousins squeezed into the back seat. Wong was the smallest so he crawled along the front seat bench, and settled down squashed next to his mum.

When they got there, Apak parked his car inside the botanical gardens. Wong's family untwisted themselves and climbed out. They followed a small creek around the looping path and watched a mother long-tail macaque, which is a type of monkey, swinging above them with her baby hanging tightly on her back.

As they rounded a corner at the far end of the botanical gardens, Wong saw a square cage with thick metal bars. He approached it with rising excitement. He loved animals, especially his fish, cats and turtles. Do you love animals too?

Wong especially loved watching unusual animals and he wondered what was inside the cage. As he got closer, a flash of orange grabbed his attention. A rope swing hung across it like a trapeze at the circus. Something orange was swinging. It looked like a scrawny version of him. It had longer arms and much more hair, but a human-like **quizzical** expression. The sign said it was a **juvenile** female orangutan. Juvenile means young, and orangutan means 'man of the forest' in the Malay language. It seemed like a good description because it looked a lot like a human. He wished he could swing like that.

Do you look this sleepy in the morning?

As he watched, the orangutan turned upside down, hanging from its legs, then stopped and stared straight at him. Wong giggled, and so did all the other people who were watching. Wong could observe the orangutan better when it wasn't moving. It had unruly hair which stuck up from its head like his did when he woke up. It had a little nose with wide nostrils, and black button eyes surrounded by a circle of pink skin. Its nose and mouth were also pink. The rest of its face was covered with soft fur the colour of rust.

Wong was then close enough to see the orangutan's hands and he held his up to compare. The orangutan had four fingers and a thumb on each hand, just like his. Its fingernails were just the same too… but his were cleaner! Unlike his, the back of its hand was very hairy.

The orangutan raised its arms up and grasped the branch. Its arms were even longer than he thought, and they looked really strong. In an instant, the orangutan turned right way up, and its feet dangled. They looked more like Wong's hands than his feet. The toes were longer than his. Perhaps, Wong thought, that was how the orangutan gripped the branches with its feet?

Do orangutan
hands look
like yours?

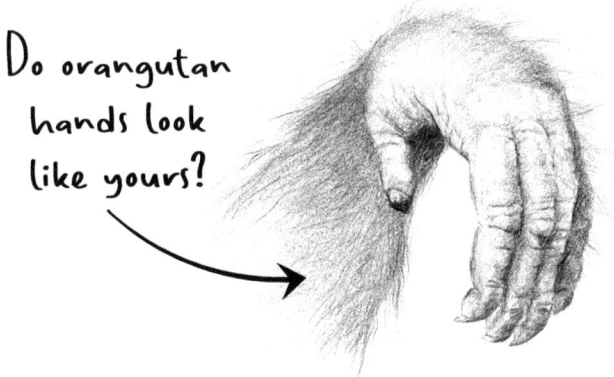

Working with animals

As Wong grew, so did his interest in unusual animals. Over time, he had a **menagerie** (which means a strange collection of animals). It included fish kept in tanks he made with black plastic, cats, dogs, turtles, breeding birds, and even a rescued common palm civet. That's a type of wild animal a little bigger than a cat with black and white stripes and dots over its fur.

Wong's favourite time was when he was with his animals. He dreamt of owning a pet shop when he was old enough, or maybe he could be an animal doctor or **veterinarian** and look after animals. Have you ever thought you would like to do that?

There are many different ways to work with animals but, unfortunately, Wong's grades at school were not quite good enough for him to be accepted into veterinarian university. Wong was very disappointed, but his brother Ben offered him hope. A year before, Ben had moved to another country called Taiwan to go to university. Even though he was scared to leave Ah Ji, Wong gathered his courage. He applied for university in Taiwan to study veterinary science and **animal husbandry**, which means looking after farm animals. When he was accepted, Wong was **ecstatic** (or VERY excited). He packed his bag, gave his mum a big hug, and boarded a plane.

Taiwan is very close to China

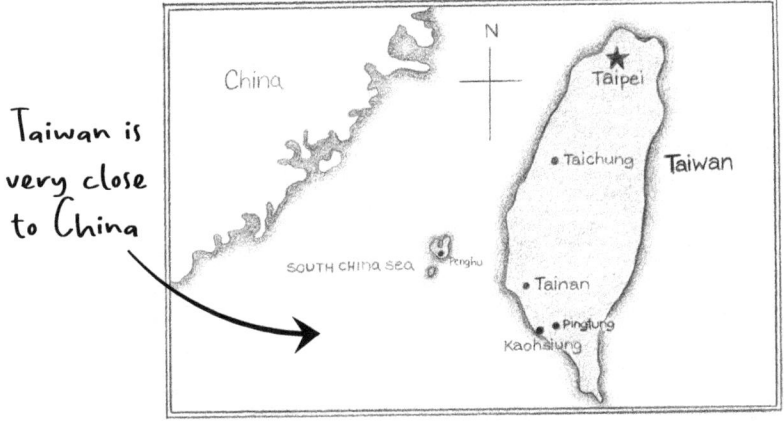

Do you think you would be homesick if you went to university in another country? Wong was, but he worked really hard, and he enjoyed what he was learning. He knew sometimes you have to give up one thing you love so you can do another thing you love.

Wong's favourite teacher

Do you have a favourite teacher? Wong's favourite teacher at university was called Professor Kurtis Pei (his last name sounds like 'pay'). Kurtis was a **biologist** (which is someone who studies animals and plants). He was also a **wildlife ecologist** (which is someone who is interested in the lives of wild animals). Wong admired Kurtis and learnt a great deal from him. Kurtis taught Wong about illegal **wildlife trafficking** (which means transporting and selling wild animals against the law), and this made Wong both angry and sad. Even though it was illegal, some people in Taiwan liked to have really unusual pets, so other people made a lot of money by catching unusual animals around the world and sending them to Taiwan. Some people even *collected* animals like you might collect Pokémon cards or stamps!

Kurtis told Wong one of the most prized animals was an orangutan. When he heard this, Wong remembered the orangutans in the botanical gardens. He knew they didn't belong in captivity, and they also didn't belong in Taiwan. Just like him, he thought, the **trafficked** orangutans must be homesick.

The rescue centre

One day, the government rescued a poor trafficked orangutan called Polly. They couldn't return Polly to the rainforest in Borneo because she was **habituated**, which means she no longer knew how to be a wild orangutan. They asked Wong's teacher Kurtis if he could keep her in one of the large cages on the eastern side of the university. He agreed.

Kurtis had a big, generous heart and before long, he had agreed to help many other animals. One day, Kurtis arrived to work to find a cardboard box dropped off at the front gate. He folded back the flaps to find a large tortoise inside. Kurtis shook his head. He wished people would stop buying wild animals! It's a lot of work taking care of so many animals, and Kurtis still had to teach at the university, so he asked Wong to be his assistant. Wong's dream to be an animal keeper had come true.

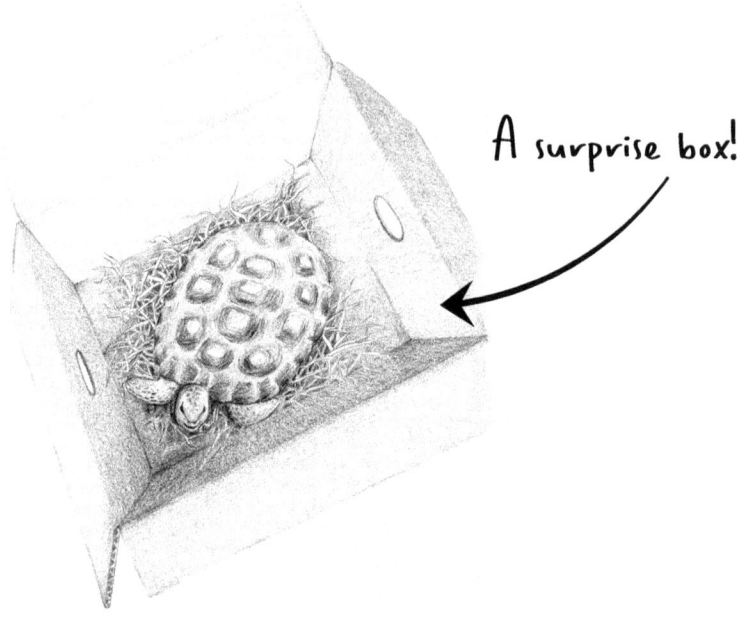

A surprise box!

That was how the Pintung Rescue Center for Endangered Wild Animals began and it is still there today. The rescue centre grew bigger and bigger. Eventually it was moved to the other side of the campus and became home to many more **endangered** animals. They have included 16 tigers who were abandoned by a breeder, two bears who were mistakenly trapped, and a gibbon which was thrown out of a housing complex because it was too noisy. If you haven't ever heard a gibbon booming, perhaps you can look it up online!

When Wong met Polly, he didn't know everything that had happened to her, but he heard so many sad stories about rescued orangutans so, as he prepared her food, he pieced them together like a jigsaw puzzle and imagined Polly's experience. It went something like this…

Polly's story

When Polly was very young, she had been minding her own business, hanging onto her mother's wiry hair high in a rainforest fig tree. She heard a strange sound. It was a chain saw, but she didn't know that. The tree started to shake so violently that her mother couldn't hang on. Polly clung on tight as they tumbled through the leaves, breaking branches as they fell. Polly's mum landed hard on her back, softening Polly's fall. Polly tried to shake her mum, but she wouldn't move. Then Polly was yanked off her mother and pushed head-first into a small wooden cage. She didn't know it, but she would never see her mother again.

Polly was scared of the chain saw

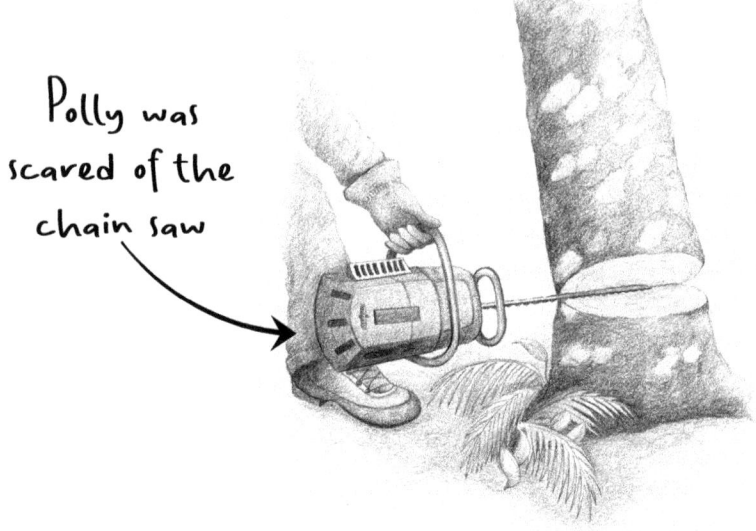

A few days later, Polly's cage was lifted into the dark, smelly inside of a fishing boat. She was scared and hungry. She was thirsty and longed for her mother's milk. She could hear a continual putt-putt sound of the engine and she held tight to the cage bars as her world began to rock. Polly felt seasick and because it was always dark, she lost track of time. She didn't know how long it had been before she was lifted out into the daylight again, but the brightness stung her eyes. The smell of salt air tickled her nostrils, and she could hear flocks of strange birds calling. Her cage was passed between hands across the water to another boat, and again her world went dark.

Eventually, the new boat stopped rocking, but Polly was barely alive.

Have you ever seen a cute puppy in a pet shop and wanted to take it home? Did you know anything about how it got to the pet shop? Wong knew that the family who bought Polly wouldn't have known the details about her horrible history. The little boy had probably wanted a baby orangutan just like the one on TV ever since he could remember, and he was really excited. He loved Polly. She was intelligent and funny. He taught her games and

Polly followed him everywhere. It was almost like having a baby sister. Polly even wore a nappy, or diaper, like a baby sister! For a year or two, all was good.

Let's face it, adult dogs aren't as cute as puppies… and adult orangutans aren't as cute as baby orangutans. As they get bigger, orangutans grow very strong, and sometimes they are dangerous. Wong thought Polly didn't like being in her cage and when she was allowed out, she started breaking things. Her family felt sad, but they couldn't let her out of her cage anymore. They didn't know what to do. At that time, it was legal to own an orangutan in Taiwan, so they tried giving her away to friends, but no-one wanted her. They considered letting her go in the forest, but their house was in the middle of the capital city, Taipei (which sounds like tie-pay), and the forest was a long way off. They were desperate.

In the middle of the night, while his children were asleep, the dad put Polly into her cage and lifted it into the car. He drove to a quiet street and opened the door. Polly climbed out.

Wong wondered how Polly would have felt when her human dad climbed back in the car and drove away.

Did she cry?

Did she try to follow him?

Was she hungry and scared?

What do YOU think Polly did?

Can you imagine how surprised the person who found Polly was?

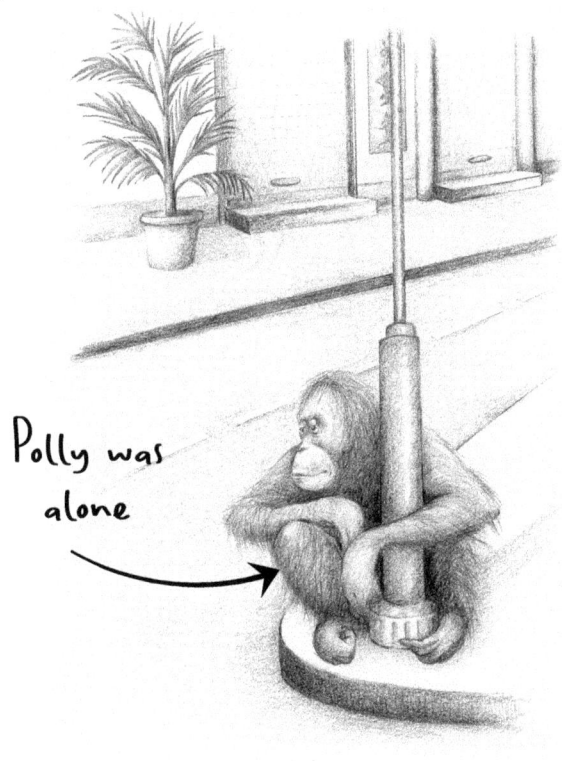

Polly was alone

Wild animals are unpredictable

When Polly was rescued, she was handed over to Wong's teacher, Kurtis, and that's how Wong came to meet her and care for her. He tried his best to make her feel safe again, but looking after animals isn't just about cuddles and cuteness. Sometimes Wong would be up all night sitting by a sick animal. Sometimes he would arrive home exhausted and stinking of animal urine from cleaning out cages. Sometimes he would spend all day building branch swings and all evening trying to dig out wooden splinters from his hands. But the more Wong worked with wild animals, the more he wanted to. Wong's family was so far away, that the animals became his family, and sometimes he even forgot they were wild.

It was a dangerous mistake.

One hot evening, Wong was feeding the animals. He took his shirt off to splash himself with water. A young, friendly rescued male orangutan reached between the bars and snatched Wong's shirt from the floor. He was pleased with himself for **pilfering** (or stealing) a new toy. Wong wasn't worried. The orangutan was like a **mischievous** teenager, but Wong didn't think he was dangerous.

Both Wong and the orangutans were homesick

PHOTOGRAPH: © WONG SIEW TE

Wong unlocked the cage door and went in to retrieve his shirt. The orangutan clung to it tightly. He was not giving up his prize without a fight. He charged at Wong as fast as he could. Wong turned to run, but the orangutan clawed at his jeans. Wong fell over and the orangutan grabbed his foot in its mouth. It squeezed its jaws shut and pain shot up Wong's leg. Wong hit back, but his strength was no match for the orangutan. The orangutan chewed further up Wong's leg. Wong screamed. On the other side of the cages, another animal keeper ran to help. Together they hit the orangutan, and finally he let go.

Wong was taken to two hospitals that day. There wasn't much blood, but his leg was crushed and bruised. It took a long time to heal and he can still see the scars, 25 years later. Wong had learnt that wild animals were unpredictable (which means you can never be certain what they will do). It was a lesson he never forgot.

Wild AND free

The next time Wong met an angry orangutan was ten years later. He was deep in the jungle of Borneo with his two assistants, Dominic and Remy. They were trying to find sun bears, who share the rainforest with orangutans and many other amazing creatures. Sun bears are **elusive**, which means they are very hard to find. They are *so* elusive, that Wong had only managed to find three bears and he had been looking for over a year. Can you imagine that?

Wong caught bears to put tracking collars on them

PHOTOGRAPH: © WONG SIEW TE

One day, Wong saw two orangutans high in the trees. He lifted his binoculars to his eyes and zoomed in to watch them. Most orangutans spend their time high in the trees because it is safe up there, and they can travel quickly swinging with their arms. They seldom bother humans, so Wong took a gulp of water from his flask and kept looking for bears.

Wong was so busy looking for signs of sun bears — like crushed termite logs or claw scratches on the tree trunks — that he didn't see the huge male orangutan in front of him. Male orangutans are heavier than females and sometimes the trees are not strong enough to hold them, so they are forced to walk along the ground.

Wong was startled when he heard a funny noise. It sounded like the raspberry sound you make with your lips pursed shut. That sound is funny when you are trying to make a baby human laugh, but Wong didn't laugh. Instantly, he knew he had stumbled into an orangutan's **territory** (the home area of an animal) and he was being warned not to get any closer. He looked up, just as the orangutan started moving quickly in his direction. It was at least 100kg which is about as heavy as you and your best friend put together. Wong yelled to Dominic and Remy, "Run!" Luckily, they got away.

People in cages, orangutans free

Wong tracked sun bears in the rainforest for six years and, in that time, he saw many more orangutans than sun bears! If you have read *Wildlife Wong and the Sun Bear*, you know that Wong now works at the Bornean Sun Bear Conservation Centre in a place called Sepilok, deep in the rainforest. This is orangutan and sun bear **habitat** (the place they naturally live).

Right next door to the Sun Bear Centre is the Sepilok Orangutan Rehabilitation Centre where they raise rescued orangutans. The part of the forest where the rescued sun bears live is surrounded by an electric fence to keep them away from people, but these orangutans are not as dangerous as wild orangutans because they have lived part of their lives with humans, so most of the orangutans roam wherever they want.

In fact, everything in Sepilok is in a cage except the orangutans. Yes, that's right! The air conditioner is in a cage because orangutans can undo screws with their fingernails; the water tank is in a cage so they don't lift the lid and climb in; and the top of every tap is removed so they don't play with the water and let it run all day long.

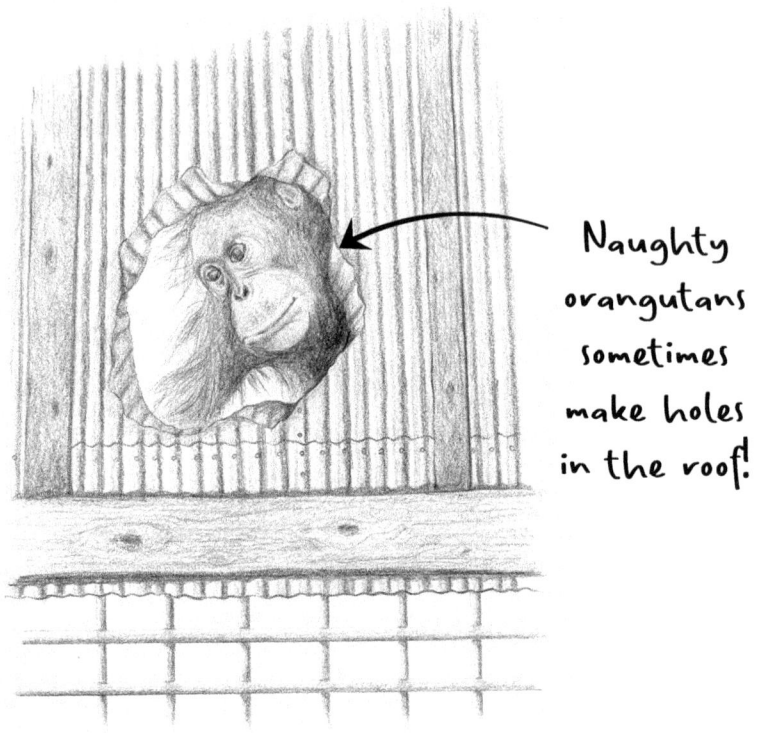

Naughty orangutans sometimes make holes in the roof!

Just like you and your friends eat lunch together in the playground at school, some of the orangutans gather on a platform in the rainforest at lunch time and wait for the keeper to arrive with a basket of bananas. Tourists pay to see them eating, and the money they pay helps buy the orangutan's food.

It was Boxing Day, the day after Christmas, when I first visited the orangutan centre. My daughter Amber (who was about your age) and I were on holiday in Borneo. It rains a lot in the rainforest, which is how it got its name. It was raining that day. Our hair was dripping and we had our coats zipped up to our chins waiting for the orangutans to arrive. Eventually, two orangutans **reluctantly** showed up – one male and one female. They weren't going to miss out on lunch, even if they got wet. As we watched, the large male orangutan reached for a tree branch. He broke off a twig of broad leaves and **fashioned** an umbrella. He then sat underneath it eating a fistful of bananas and staring at us. I glanced around at the other humans huddled in a heap and asked Amber "Who do you think is watching who?"

This orangutan made an umbrella out of leaves

The day after that, I met Wong. We became friends and, since then, I have spent a lot of time in Sepilok working with Wong. Everything you do in Sepilok has to try to outsmart orangutans. It took me a long time to get used to winding up the car windows tight, and locking all the doors so the orangutans couldn't get in. During the day, clever orangutans watch every move that humans make. They remember every mistake or opportunity for fun.

One day, we were having a meeting in Wong's office. Someone had left the glass sliding door unlocked and an orangutan tried to come into the meeting! Can you imagine that? Another day Ronny the maintenance man was washing the stainless-steel rail along the walkway. Wong called him on his walkie talkie and asked for help, so Ronny left his bucket and sponge to walk back to the office. When he returned, an orangutan was washing the rail for him!

The bear keepers collect old rice sacks. They are big enough for you to get inside, and they use them to to collect leaves in the forest so they can make **enrichments** for the bears. If they leave an empty rice sack for a moment, a lazy orangutan will grab it. Then, after 5pm when all the people have gone home, it will drag the rice sack up to the roof and use it as a sleeping bag so it doesn't have to build a nest. Yes, orangutans build nests... I'll tell you more about that later.

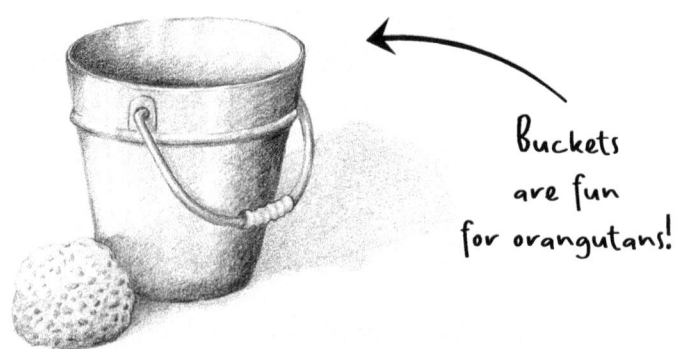

Buckets are fun for orangutans!

There are many other funny stories. But the funniest story was when Wong got locked into his office.

The price of freedom

A few years ago, Wong's office was getting too small for him and his growing sun bear team. While builders made it bigger, the staff at the orangutan centre said he could use the building next door where visiting vets sometimes stayed. Wong packed up all his office things into boxes and carried them up the stairs towards his temporary office. On the balcony, he glanced over to the cage where a naughty orangutan was having forced time out for stealing a tourist's camera. Another orangutan was sitting at the other end of the deck. Wong thought it looked like she was waiting for her friend to get out of detention. Wong put down his box and rested his foot against it while he quickly unlocked the metal grate in front of the door to let himself in. All the while, he kept one eye on the free orangutan.

Chilling on the deck

PHOTOGRAPH: © SEN NATHAN

Being in charge of an entire rescue centre is a big job, and Wong works very hard. Sometimes he forgets to go home until late in the afternoon when the sun is setting and the rainforest colours have changed from green to orange. One day, Wong waved to each of his team as they packed up their lunch boxes and said goodnight, but he just needed to finish off writing his emails. Eventually, he turned off the light and headed for the door. But there was a problem. The metal grate was locked. Standing on the other side of the bars, just like a person visiting an animal in a zoo, was a large orangutan. Its arm was stretched out and the door key dangled from its hand. What could Wong do?

First, he tried talking nicely to the orangutan, asking for the key back. But that didn't work. Next, he looked around for something he could stick between the bars to snag the key. There was nothing around.

"Maybe I could trade the key for food?" he thought.

Wong went back into the office, and into the kitchen. He opened the fridge. Hopefully someone had left a slice of cake or a juicy mango! His heart dropped when he opened the door. The fridge was empty... except for two small limes. Do you think an orangutan would be happy with limes? Wong wasn't sure either.

What would
you do?

Wong went back to the front door. The orangutan was still there. He held out one of the limes and tried to trade. The orangutan took the lime. It raised it up to sniff, keeping tight hold of the key in the other hand.

Wong's plan hadn't worked.

Wong held his breath. He had one chance left. **Gingerly** (which means carefully), he stretched his arm out with the last lime on his palm. His other fingers were crossed.

The orangutan looked at Wong.

It looked at the lime it was holding.

It looked at the other lime in Wong's hand.

It looked at the key.

For Wong, seconds felt like minutes. Then the orangutan offered Wong the key, trading it for the lime. Wong sighed in relief. Both Wong and the orangutan thought they had bargained well. As the orangutan climbed a tree with his prize, Wong unlocked the door and went home.

Want to know more about orangutans?

Did you enjoy that story? Now, why don't we learn a bit more about orangutans. THEN, it's your turn to become a scientist!

The primate family

If you have joined the Wildlife Wong Kids' Club and downloaded your free Nature Journal (www.sarahrpye.com), you will already know what a mammal is. If not, let's recap... a mammal is an animal with a backbone. It has warm blood and hair. It sweats when it's hot, drinks its mother's milk as a baby, and is born alive (not in an egg). Yes, that's right, you are a mammal!

Mammals are divided into other groups based on their similarities, and one of those groups is called primates. There are three hundred different species of primate in the world. They include animals like monkeys, apes and humans. These are organised into different families and one of those families is called Great Apes, or **Hominidae**. That family includes gorillas, orangutans, chimpanzees and bonobos (which are very similar-looking to chimpanzees, but a little smaller). Humans are also in the Hominidae family.

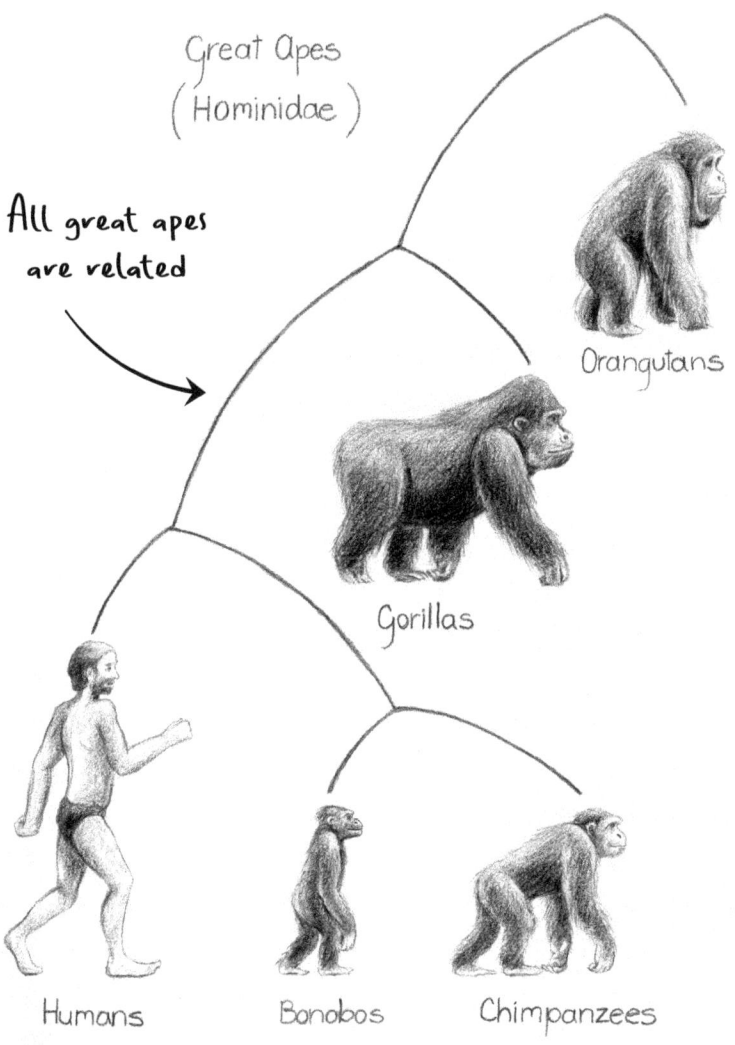

Great Apes
(Hominidae)

All great apes
are related

Orangutans

Gorillas

Humans

Bonobos

Chimpanzees

Imagine you have a pile of Lego blocks and you try to build a human. It would be pretty hard without an instruction book. DNA is like an instruction book for building each animal. Scientists study DNA, to find out which of these animals is our closest relative and it turns out chimpanzees and bonobos are the closest to human. We actually share almost 99% of our DNA, which means the instruction booklet for a human is almost the same as the booklet for a chimp. Orangutans are the next closest. We are 97% the same as orangutans.

What do orangutans look like?

When orangutans are born, they look like hairy human infants with longer arms. Their faces look a bit like the cartoon Homer Simpson! Orangutans only have one baby at a time and the baby stays with their mum for the first 6-8 years of their life, holding onto her fur as she swings between the branches in the rainforest. This is an important time because they learn skills from their mum.

Baby orangutans learn from their mum

As they grow up, males start to look different from females. When they are fully grown, at about 15 years old, males weigh about twice as much as females. When males reach adulthood, or **maturity**, they develop large cheek pads which make their face look less human. Some even look round and flat, like a dinner plate! Female orangutans think it looks very attractive.

Orangutans have very long arms and their hands reach the ground as they walk. They have **opposable thumbs,** just like humans. This means their thumb is opposite their fingers.

This helps them grab branches and swing in the trees. Scientists say orangutans are about seven times as strong as humans. They can lift as much as 250kg. That's about the weight of an adult pig. Can you imagine that?

Orangutans live to be about 40 years old in the wild. In captivity, they can live up to 70 years because there are less dangers in zoos.

Adult males look different from females

What do orangutans eat?

Orangutans eat mostly plants, but sometimes they eat other living things too. The word for animals with this sort of diet is **omnivore** (which sounds like om-nee-vor). **Omni** means 'all things' and **vore** means 'animal that eats'. However, more than half (about 60%) of what orangutans eat is fruit. They love tropical fruits like wild longan (which looks a bit like a lychee) mangos and figs. Do you have a favourite fruit? Is it the same as an orangutan? They also eat **succulent** (which sounds like suck-you-lent and means juicy) young leaves, colourful flowers and plant shoots. They even eat soil and tree bark! When they eat other creatures, it is usually termites, other insects and small vertebrates (which is an animal with a backbone). They also sometimes eat bird eggs if they find a bird nest.

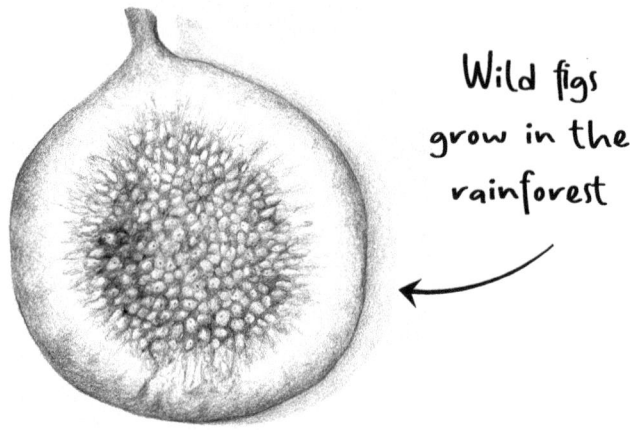

Wild figs grow in the rainforest

Where do orangutans sleep?

Orangutans are pretty amazing because they make nests to sleep in. First orangutans weave branches into a cradle high in the trees. Then they line it with broken sticks to make a mattress. Lastly, they make it comfortable with soft leaves. In the wild, orangutans make a new nest every night and sometimes they even make a second one to have a nap during the day. Can you imagine if you had to make a new house every day?

Why are orangutans important?

Orangutans have become the celebrities of the rainforest. Just like celebrities, people travel from all over the world to see them and sometimes people become interested in saving the rainforest because they love orangutans. The first time I went to Borneo it was because I wanted to see orangutans and I love the rainforest even more now. Orangutans have been called an **umbrella species**. That doesn't mean they make umbrellas (even though they do!). It means if we protect orangutans and their home, we indirectly protect many other species — even some of the ugly ones!

The importance of poo

I bet you haven't ever thought about how important poo is. When you eat, some of your food which you don't digest ends up in your poo. YOUR poo probably travels down the toilet to the sewage farm where it is treated, but orangutans' poo is important to the rainforest.

Orangutan poo contains heaps of fruit seeds, and by pooing all over the place, orangutans help new trees grow. The new trees eventually grow fruit which orangutans eat. This is called a **symbiotic** relationship, which means trees and orangutans work together to make the world better for both of them. Can you think of other species that have symbiotic relationships with each other?

What happens if the trees don't grow?

No matter how quickly orangutans are pooing they are still in danger of losing their home. New rainforest trees are not growing fast enough, while people keep cutting old trees down. We call dead trees wood, or timber. The timber industry is important for building homes, furniture, and even for making books. Can you list five things in your house made of wood?

Every single one of those used to be a living tree. In many cases it's better to use wood than plastic, because it is **biodegradable** (which means when it breaks down it is still used in nature). When humans plant new trees to replace all the trees we cut down, then the timber industry is **sustainable**, which means it can basically go on forever without being a problem for the environment. But, when we don't plant enough new trees, or we don't give them enough time to grow, then the forests get smaller and smaller. That means the orangutan's home gets smaller, and there is less food for them to eat.

But humans need to eat too...

Sometimes the rainforest in Borneo is cut down to clear the land and grow other plants for humans to use. Oil palms are one of these **replacement** plants. You can probably guess what we get from oil palms? Yes, that's right, oil! Oil palms grow very quickly and they produce a lot of oil which is used in all kinds of things like chocolate, cookies, instant noodles, and even shampoo. In some countries, food packages need to say whether the food contains palm oil. If you are in one of those countries, you can choose between products with palm oil, or products without palm oil. In Australia, we don't have that law which makes it harder to avoid using palm oil.

Can you see where the palm oil plantation meets the rainforest?

PHOTOGRAPH: © GLENN HUROWITZ

Sometimes rainforest is burnt down to clear land and grow human food like corn, rice, cows (beef) or pigs (pork). The fires cause air pollution, and cows and pigs release a gas called **methane** when they poo which warms the air and adds to climate change. Some people have reduced how much meat they eat to help save the rainforest.

Experts think there will be no more orangutans on earth by the time you are as old as me. That's pretty scary. Scientists are trying their best to save animals from extinction. Being a scientist, like Wong, is a very important job, don't you think? But even those of us who are not scientists can help by changing the way we live. Then, perhaps we can prove the experts wrong and keep orangutans (and the rainforest!).

Next time, I will tell you all about *Wildlife Wong and the Pygmy Elephants*, but if you haven't yet read *Wildlife Wong and the Sun Bear*, perhaps you want to read that one next.

First, do you want to do some science experiments?

Experiment 1: Punny Pye!

Do you know what a pun is?

It's a joke that's funny because a word has two meanings. Probably some of the jokes that make you cringe are based on puns. For instance, "What are the pie rates of the Caribbean? ... $2.50 in Jamaica, and $3.00 in Tortola!"

I grew up in the Caribbean. That's a tale for another day but, when I lived there, my favourite dessert was called Key Lime Pie. 'Key limes' are small, sweet limes that turn yellow when they are ripe. If you think back to the story of Wong and the orangutan who stole the key, I reckon Key Lime Pie would be the orangutan at Sepilok's favourite dessert. Do you get the pun? Can you see another pun in the experiment title?

In chemistry (which is a type of science), you start with special equipment, measurements and instructions. Then you use scientific chemical reactions and observation to create something new. Cooking uses all the same things, so today our chemistry experiment is making Key Lime Pie!

You may need an adult to help you with this one, but let's get started.

Key Lime Pie

This recipe makes eight generous, delicious slices.

Equipment

25 cm (10 inch) pie dish
Rolling pin
Strong plastic bag
Large spoon for mixing
Tablespoon
Teaspoon
Measuring cups
3 small bowls
I large bowl
Egg whisk
Grater
Oven

Ingredients

For the crust, you will need:

- 180 grams (1.5 cups) of digestive biscuits (in Australia or the UK) or graham cracker biscuits (in North America). If you are gluten intolerant, you can use gluten-free biscuits instead.
- 70 grams (5 tablespoons) of melted, unsalted butter

For the filling, you will need:

- Zest from 3-4 limes
- More limes. You need 175 ml (3/4 cup) of lime juice (they don't have to be key limes, you can use any limes)
- 400 ml (usually one can) of sweetened condensed milk
- 5 large eggs

For the topping, you will need:

- 300 ml whipping cream
- 1 teaspoon of icing sugar
- The rest of the lime zest

Steps:

Make the crust:

1. Preheat the oven to 180 degrees Celsius or 350 degrees Fahrenheit.
2. Put the biscuits in a strong plastic bag and crush them into crumbs using a rolling pin. This is really satisfying, but if you have a food processor, you can use that instead!
3. Put the crumbs in a large bowl and add the melted butter.
4. Mix with the spoon, and then your hands, until the crumbs start sticking together. Add a little more butter if needed.
5. Scoop the mixture into the pie dish and press it down so it covers the bottom of the plate and the sides evenly.

6. Bake the pie crust for 10 minutes. Take it out and let it cool down.

7. Don't turn off the oven... you need it again later.

Make the filling:

1. Grate the rind of four limes until you have about two tablespoons full. Be careful not to grate your fingers!

2. Cut the limes in half and squeeze the juice out until you have a cup or 240ml. This might take more than four limes. Take out any pips that ended up in the juice!

3. Add half of the lemon zest. Put the other half aside for later.

4. Use two small bowls to separate the egg yokes from the egg whites. If you are super clever, you can use the egg shell to do this, but you can also use you hand with your fingers a little apart. (Check out the video in the Wildlife Wong Kids Club www.sarahrpye.com).

5. Put the egg white back in the fridge. (It can be used for another experiment or meal!)
6. Add the lime juice and zest mixture to the egg yolks.
7. Add the sweetened condensed milk to the lime juice mixture.
8. Whisk the mixture until it is all one colour (with lime specks in it from the zest).
9. Pour the mixture into the cooled pie crust and smooth it out with a spoon.
10. Bake for about 15-20 minutes. It is done when the top of the pie isn't liquid anymore, but the pie is still jiggly.
11. Take the pie out of the oven and cool it in the kitchen for two hours.
12. When it is cool, put it in the fridge for at least 4 hours (or overnight). This is the hardest part because you will probably want to eat it!

Make the topping:

1. Just before you are ready to eat the pie, put the cream into a small bowl.
2. Add the icing sugar and whisk it until it is thick. (if you have an electric whisk you can use it, but an ordinary whisk is fine.)
3. Use a spoon or flat knife to spread the cream over the pie. It doesn't have to be flat... if you want you can make it look like cream mountains!
4. Sprinkle the rest of the lime zest on top, like snow.
5. NOW you can eat it!

Time to
tuck in!

How does it work?

Key Lime Pie is a type of custard pie and it's pretty cool how the middle changes from a runny liquid to a jelly-like solid.

Custard thickens because it includes eggs. Eggs contain a type of protein that **coagulates** when it is heated (which means it turns from liquid to solid). That's why eggs turn hard when you boil or fry them. The heat does something called **denaturing** the protein, which means it breaks down the structure. If you looked at the eggs with a microscope, you would see long chains of amino acids.

Some amino acids like water, and some don't. The ones that are scared of water (**hydrophobic**) panic when it gets hot. They rush around looking for each other for protection. The chains join together against the danger and this forms a solid mass. There's a problem though. If you heat up the mixture too quickly, the hydrophobic amino acids gather really quickly, causing lumps in the custard. Sugar is the antidote to this problem. Hydrophobic amino acid bounces off sugar, which means it takes more time for the acids to gang up, and there is less chance of **curdling** (which is what it is called when proteins create lumps). In this recipe, the sugar is in the sweetened condensed milk.

Experiment 2: Make a nest

Can you build a nest like an orangutan? They bend branches and weaves them together with their hands and teeth and it is not as easy as it looks. Unless you live in Borneo, I am pretty sure you won't be able to find an orangutan nest to copy, so you will have to get creative for this experiment. You also don't live in the trees, so let's see if you can make a nest on the ground instead.

Orangutan nests are hard to see!

PHOTOGRAPH: © BARBARA KATSIFOLIS

Trees need their limbs just like you need yours, so don't break off tree branches. Instead, gather things that are already on the ground, or ask someone who likes gardening if you can have their garden waste.

You will need:

As much natural nest material as you can:

- leaves
- trimmed branches
- palm fronds
- weed vines
- grass cuttings
- sticks
- mud

Steps:

1. Lay your nest materials out on the ground and sort them into sizes, shapes and softness.

2. Use the most **malleable** sticks first. Malleable means they bend easily. Palm fronds work really well if you have them!

Bend them into a cradle and weave them together so the sides come up off the ground.

3. You can use vines to tie it together, or mud as glue.

4. Make sure the circle is big enough for you to get into.

5. When the cradle is secure, use broken sticks to create a mattress inside the cradle just like an orangutan does.

6. Now line your nest with the softest leaves you have to make a pillow.

7. If you have any materials left over, strengthen the nest by weaving them into the outside.

8. You can even decorate your nest with colourful leaves and flowers!

9. When you are finished, see if you can move your nest without it falling apart... if you can, well done!

10. Now climb in and see if your nest is comfortable. Do you think you would like to sleep in it?

Check out the
video in the
Wildlife Wong Kids'
Club to see how
Aysha and Levi
made this nest!

New words

Some of the words or phrases in this book are bold. Here's what they mean. They are in alphabetical order. If a word (or phrase) starts with A, AN or THE, it is a noun (a person, place or thing). If it starts with TO BE, it is a verb (a doing word). An adverb describes (or adds to) a verb, and an adjective describes (or adds to) a noun. I reckon it should be called an adnoun!

Animal husbandry — (noun). The science of breeding and caring for farm animals.

Biodegradable — an adjective which describes something which breaks down naturally over time.

A biologist — a scientist who studies animals and plants.

To coagulate — turn from liquid to solid.

Curdling — (verb) when proteins gather into lumps.

Denaturing — a verb for breaking a protein apart.

An ecologist — a person who studies how animals and plants live and work together.

To be elusive — very hard to find.

To be endangered — at risk of extinction.

Endemic — an adjective which means something which is native to a specific area.

The enrichments — toys made for animals in captivity.

To be ecstatic — very excited.

To be fashioned — made into something.

Gingerly — an adverb which means carefully.

A habitat — the type of place where an animal lives.

To be habituated — to be accustomed, or used to something.

The Hominidae — the family of great apes.

Hydrophobic — an adjective which describes something scared of water.

Introduced — an adjective which describes a species which doesn't live there naturally.

Juvenile — adjective which describes a young animal, but older than a baby.

To be malleable — bends easily.

Maturity — (noun) after childhood.

A menagerie — a strange collection of live animals.

Methane — A type of flammable gas which can warm the atmosphere.

Mischievous — adjective which describes someone who is fun or makes trouble.

An omnivore — an animal that eats both plants and other animals.

Opposable thumbs — thumbs (which are nouns) opposite to fingers so they can grip.

A peninsular — land that juts out into the ocean.

To be pilfering — stealing.

Quizzical — adjective which describes a puzzled facial expression.

Reluctantly — an adverb which describes doing something you don't want to do.

A replacement — something which has taken the place of another thing.

To be splayed — spread out.

Succulent — an adjective which means something is juicy.

Sustainable — an adjective which means something can basically go on forever without being a problem for the environment.

Symbiotic — adjective which describes things that work together to make the world better for both of them.

A tailor — someone who sews clothes.

A territory — the home area of an animal.

To be trafficked — to be traded, or sold illegally.

An umbrella species — an important species that, if saved, would also save many other species.

Unpredictable — an adjective which describes something that changes without warning.

A veterinarian — animal doctor.

A wildlife ecologist — someone who studies the lives of wild animals.

Wildlife trafficking — illegally trading or selling wildlife.

Do you want to read more?

There are many different animals in Borneo and Wong has stories about all of them! Why not check out these ones next?

Do you want to help rainforest animals?

Here are a few ideas:

- Lend this book to your friends so they can learn about orangutans too
- Do a school project on orangutans
- Adopt an orangutan or a sun bear with you family or your class at www.orangutan-appeal.org.uk or www.bsbcc.my
- Join the Wildlife Wong Kids' Club and download your free Nature Journal at www.sarahrpye.com
- If you waste less and buy less, less rainforest needs to be cut down
- If you want to avoid using unsustainable palm oil, check out a list of products online

- Volunteer with a conservation group in your own area – the entire environment needs help, not just Borneo!
- If you are old enough, connect with me on Facebook or Instagram
- Email me your review of this book. If it is written well, I might ask you to write one for the next book, and you will get your picture in the front!

For teachers and parents:

Sarah Pye is available for speaking engage-ments, keynote addresses and hands-on workshops online and in person. For more information visit:

⊕ **www.sarahrpye.com**

This book was printed on demand (POD) which reduces waste and saves our trees.